Shapes Everywhere

Shapes on a Farm

Oona Gaarder-Juntti
Consulting Editor, Diane Craig, M.A./Reading Specialist

A Division of ABDO
ABDO
Publishing Company

visit us at www.abdopublishing.com

Published by ABDO Publishing Company, a division of ABDO, P.O. Box 398166, Minneapolis, Minnesota 55439. Copyright © 2014 by Abdo Consulting Group, Inc. International copyrights reserved in all countries. No part of this book may be reproduced in any form without written permission from the publisher. Super SandCastle™ is a trademark and logo of ABDO Publishing Company.

Printed in the United States of America, North Mankato, Minnesota
062013
012014

Editor: Liz Salzmann
Content Developer: Nancy Tuminelly
Cover and Interior Design and Production: Oona Gaarder-Juntti, Mighty Media, Inc.
Photo Credits: Brand X Pictures, Comstock, Creatas Images, Hemera Technologies, Jupiterimages, Shutterstock, Stockbyte, Thinkstock

Library of Congress Cataloging-in-Publication Data
Gaarder-Juntti, Oona, 1979-
　Shapes on a farm / Oona Gaarder-Juntti.
　　　p. cm. -- (Shapes everywhere)
　ISBN 978-1-61783-416-5
　1. Shapes--Juvenile literature. 2. Farms--Juvenile literature. 3. Farm buildings--Juvenile literature. I. Title.
　QA445.5.G334 2013
　516'.15--dc23
　　　　　　　　　　　　2011051116

Super SandCastle™ books are created by a team of professional educators, reading specialists, and content developers around five essential components—phonemic awareness, phonics, vocabulary, text comprehension, and fluency—to assist young readers as they develop reading skills and strategies and increase their general knowledge. All books are written, reviewed, and leveled for guided reading, early reading intervention, and Accelerated Reader® programs for use in shared, guided, and independent reading and writing activities to support a balanced approach to literacy instruction.

Table of Contents

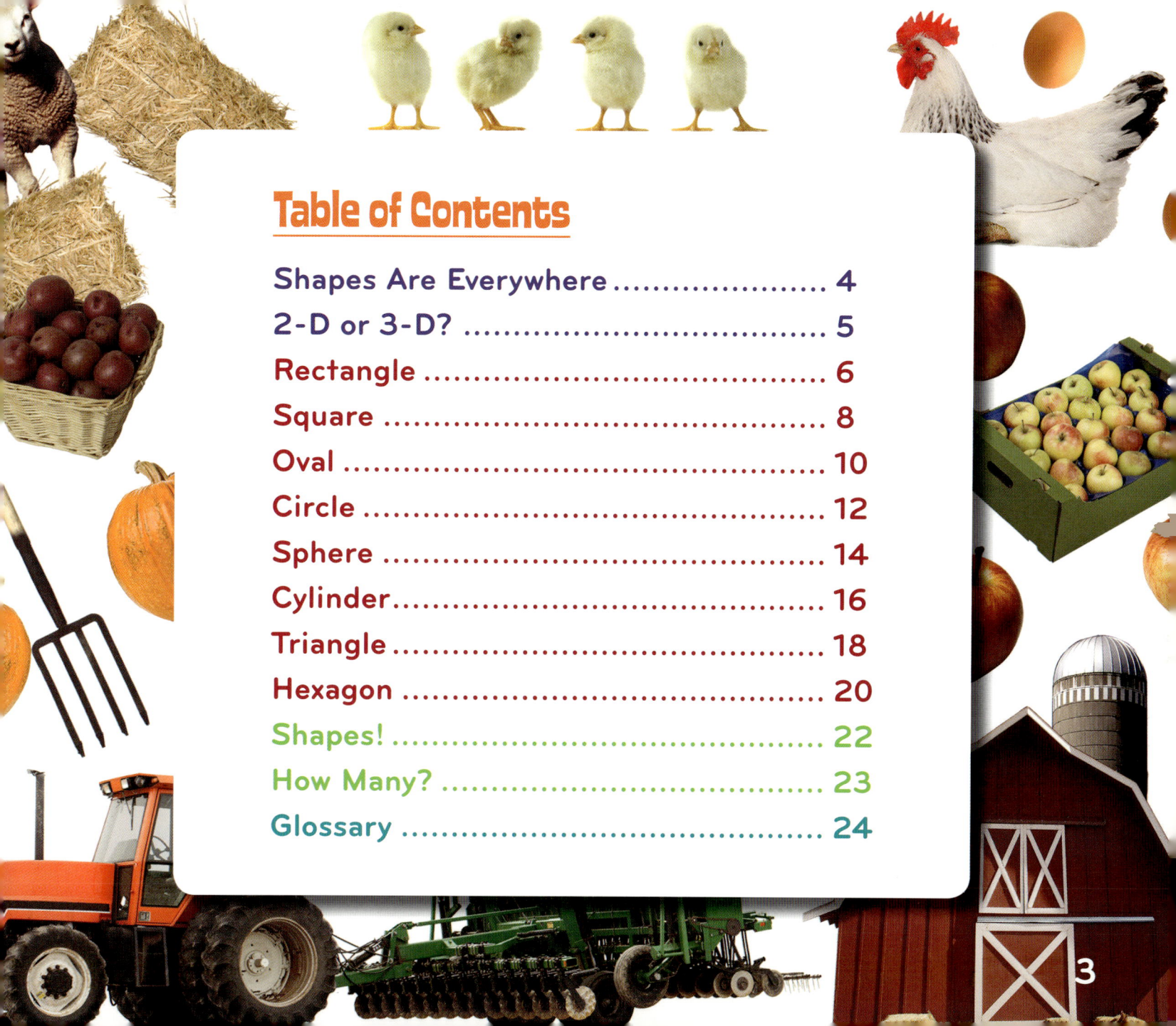

Shapes Are Everywhere 4
2-D or 3-D? 5
Rectangle 6
Square 8
Oval 10
Circle 12
Sphere 14
Cylinder 16
Triangle 18
Hexagon 20
Shapes! 22
How Many? 23
Glossary 24

Shapes Are Everywhere

Shapes are everywhere on a farm! Here are some shapes you might see. Let's learn more about shapes.

2-D or 3-D?

2-Dimensional Shapes

Some shapes are two-dimensional, or 2-D. A 2-D shape is flat. You can draw it on a piece of paper.

circle
2-D shape

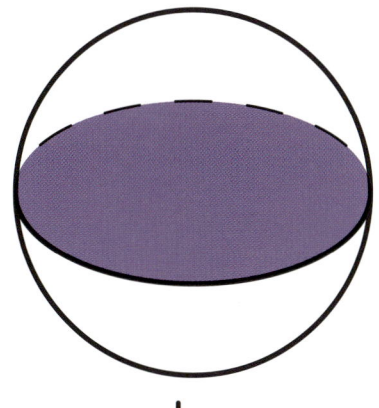

sphere
3-D shape

3-Dimensional Shapes

Some shapes are three-dimensional, or 3-D. A 3-D shape takes up space. You can hold a 3-D shape in your hands.

RECTANGLE

The door opening is a rectangle. Hannah has a new horse named Sparkle. Hannah feeds and brushes Sparkle every day.

SQUARE

The barn windows are square. Max visits his grandpa's farm. He likes to help with the **chores**. He helps move the hay in the barn.

OVAL

Eggs are oval. Susan gets the eggs from the henhouse every morning. Different kinds of chickens lay different colored eggs.

CIRCLE

The tractor wheels are circles. Henry will learn how to drive a tractor when he is older. His dad will teach him.

SPHERE

The apple is a sphere. Andrew and his family go to an apple **orchard** every fall. They get to pick the apples. Then they go on a hayride.

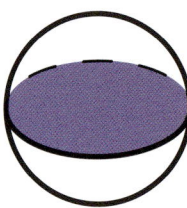

15

CYLINDER

The hay **bales** are cylinders. Mary's family drives in the country on Sundays. She sees hay in the fields outside the car window.

17

TRIANGLE

The rooster's comb has triangles on it. Charlie wakes up when he hears the rooster crow. The rooster crows very early!

HEXAGON

The **combine's** blades are in the shape of a hexagon. The combine is cutting wheat. Ella's family works hard during **harvest** season.

21

Shapes!

Here are the shapes in this book, plus a few more. Look for them when you are on a farm!

diamond

rectangle   pentagon   hexagon   octagon   square

star   heart   oval   triangle   circle

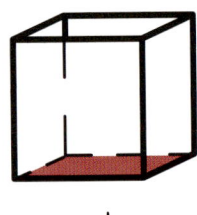 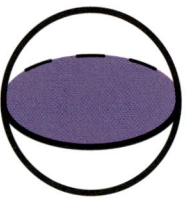  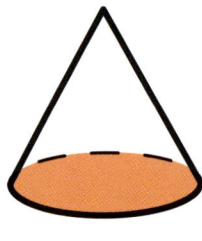 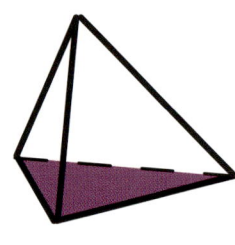

cube   sphere   cylinder   cone   pyramid

How Many?

How many shapes can you find in this picture?

23

Glossary

bale – a large bundle of something tied tightly together.

chore – a regular job or task, such as cleaning your room.

combine – a large machine that cuts and cleans grain in a field.

harvest – to gather or collect crops.

orchard – a place where fruit or nut trees are grown.